Frontispiece: The Medway Bridge

Ministry of Transport

THE APPEARANCE OF BRIDGES

Her Majesty's Stationery Office London 1969

First published 1964
Second edition 1969

Contents

Cover: Lune Bridge, Lancaster Bypass

1 Kelso Bridge *Rennie*

2 Pont-Cysyllte *Telford*

3 Clifton Suspension Bridge *Brunel*

1

2

3

Introduction

1. The recommendations given in this booklet derive mainly from consideration of views expressed by the Royal Fine Art Commission on many individual schemes submitted to the Commission over a period of years. In addition the Commission has given valuable help and advice in drafting the text.

2. The illustrations are not intended in every case to represent ideal designs, and in some cases they do no more than amplify some specific point in the text.

3. To many, some of the ideas propounded may seem obvious and elementary. But the skill and experience of designers vary greatly, and it is hoped that the indulgence of the more able will be granted for an attempt to assist those less familiar with the many problems concerning the finished appearance which arise in connection with bridge design. Even the more experienced may find some of the suggestions helpful.

4. It should be emphasized at the outset that this booklet is intended for guidance only. The last thing that either the Ministry or the Commission would wish to see is a set of standard solutions to all these problems or the automatic application of any particular formula. Variety is a vital element in the enjoyment of things seen; not only that variety which comes from differing local character and tradition, but also that rarer sort which is to be found in occasional brilliant departure from the orthodox. Britain is fortunate in possessing notable examples of such bridge designs and it would be a matter for regret if anything said here tended to sap the sources of this originality.

5. Indeed this country can boast as fine a tradition in bridge building as any in the world. Such men as Rennie, Telford and Brunel enjoyed international reputations, and their works are a source of continual admiration. Why should not our own generation, for all its differences in taste and technique, make as notable a contribution? (Plates 1, 2 and 3).

6. Of course it is only in exceptional cases that bridge building calls for a spectacular tour de force. Most of the work is bound to be on a more modest scale, but there is no reason why the humblest new bridge should not enhance the appearance of its setting. All too often in modern times it has tended to detract from it. It is not so much that standards of taste have declined, though this may be partly responsble; it is rather the lingering tendency to concentrate exclusively on structural requirements and economy in the vague and ill-founded hope that appearance will invariably look after itself. When special efforts are made the erroneous idea still occasionally persists that a satisfactory solution may be found by adding features, derived perhaps from some outmoded practice.

7. The problem becomes of increasing importance with the greater efforts now being made to bring the road system of this country up to the standards required for modern transport. Not only has provision to be made for a far greater volume of traffic but also for faster and heavier vehicles, while it becomes increasingly difficult to find room for expansion. Old bridges must be strengthened and widened or rebuilt; by-passes with their own new bridges formed round congested centres; motorways created for fast through traffic, clearing all existing roads and railways, rivers and canals.

4 King George V Bridge, Glasgow

5 Chiswick Flyover

6 Donnington Bridge, Oxford

4

5

6

General principles

8. It is generally accepted that *expression of function* is the basis of good design, that any adjustment or addition required to improve the appearance should exploit this functional basis and not run counter to it. But this is not to say that well considered departure from this principle may not on occasion be welcome. It is usually a mistake, however, to attempt deliberate deception; for example, to attach shallow stone voussoirs to a flat concrete arch. (Plate 4). There are satisfactory ways of finishing the structure without such falsity.

9. No less important is the development of *character and individuality* in design. A bridge should not only be attractive to look at, it should not only express that sense of achievement, however modest, which is inherent in all well designed bridges, it should have in addition a natural and permanent association with its setting. In other words, it should not be just 'one of those stock bridges'; it should be 'this particular bridge' carrying this road over this river, having its own individuality, although there will of course be recurring situations which justify repetition of a really good design.

10. This permanent relationship to setting calls for special attention to scale, to form and to choice of materials. A broad open landscape will call for a bolder treatment than an intimate village setting, even though the widths of road carried and river crossed may be the same in both cases. A form of bridge appropriate to a rocky gorge would be unlikely to suit a classical town centre; similar considerations apply to the materials used. In the former case a rough stone arch in the local style might provide a satisfactory answer, in the latter a smoother finish with nicely moulded edges might be called for. These should provide harmony with forms and textures that are likely to remain. In some cases, however, a better solution can be found in providing a foil or contrast. It might, for example, be much more economical to throw a light concrete structure across the gorge. In such a case it is probably better to exploit its contrasting character to the limit; an exciting contrast is usually better than a half-hearted effort at matching.

11. Where a road requires the construction of either a high embankment or a viaduct consideration should be given to the effect of each of these upon the landscape as well as their relative cost before deciding between the two alternatives.

12. Another important consideration is *proportion:* the ratio of height of opening to span, of width of structure to span, of apparent depth to span and so forth. The main dimensions of any bridge will be largely determined by functional and economic considerations, but a variation is usually possible within reasonable limits and in the choice of precise dimensions the final appearance of the bridge must be a determining factor. There can be no absolute ruling in this matter of ratios. It is the eye which judges, for example, whether the depth looks right for the span, as well as being right structurally; sometimes it may be desirable to break the face by introducing a horizontal shadow; this can be done either by setting back the lower part of the face or by introducing a chamfer. (Plate 5).

13. It is generally undesirable in a bridge of many spans to vary the length of span, except where there is an appreciable rise throughout the length, when it may be better to keep the same ratio of span to height in each case. Where there is an appreciable rise towards the centre of the bridge it is generally better to avoid a central pier. In spite of the old dictum on unresolved duality, however, there are cases, such as flat bridges over dual carriageways, where a central pier provides the most satisfactory arrangement.

14. Closely related to these problems of proportion is that of *apparent weight*. There is a general feeling nowadays, partly perhaps as a result of increased speeds, that a bridge should look as effortless as possible. The fullest exploitation of the potentialities of steel and concrete is also in itself a source of pleasure, part of the sense of achievement to which reference has already been made. (Plate 6). There may, however, be exceptional cases, for example the breaching of an old fortification for a new traffic route, where a feeling of weight is a positive advantage.

15. The best bridge designs have always had a certain *simplicity* about them. This is especially important nowadays when those who pass over or under them are usually travelling at a speed which allows them to take pleasure in the general effect of a design rather than in any intricate detail. The detailed treatment must, of course, receive careful attention, but it should be considered in relation to the general effect and any sort of 'fussiness' should be rigorously avoided.

15a. Paragraphs 10 and 11 refer to the importance of the existing landscape in determining the design of a new bridge. At the same time any new bridge is itself bound to alter the existing landscape. Often, and perhaps most of all where a new complex of flyovers and slip-roads are to be created for a road junction, this will call for special landscape work, in addition to the normal embanking and cutting, if a satisfactory relationship between structure and setting is to be achieved.

7 Rhinefield Walk Footbridge, New Forest

8 Invermoriston Bridge

7

8

Choice of type

General considerations

16. Attempt has been made above to outline some of the general principles which can guide the designer of bridges when he is considering the appearance of his work. It is now proposed to refer in greater detail to the application of these principles, and first as influencing the choice of the type of bridge to be used. At this stage the question of good appearance, both in the structure itself and in its relation to its surroundings, should not be overlooked.

17. The first essential is that the bridge should be *suitable for its purpose*, in function, in appearance and in appropriateness to its setting.

18. It is not essential to follow the same character in two bridges simply because they are carrying or crossing the same road. Attempts to achieve similarity in appearance should be avoided particularly when the functional requirements obviously call for differences in design, for example a railway bridge and a footbridge. Even with similar requirements the character of the bridges can sometimes be varied with advantage especially when the road is running through monotonous country.

19. On motorways, on the other hand, physical requirements as well as speed and economy of construction demand a considerable degree of standardisation. In such cases it is particularly important that the type design shall be really good enough to stand up to frequent repetition.

20. A common problem occurs where roads cross at different levels. A choice has to be made between the open type with say four spans (three if there is no support in the central reservation) and the bridge having solid abutments at the outer edges of the lower road. The open type is appropriate when the lower road is in wide open cut so that the spans at either end of the bridge permit the sloped sides of the cutting to continue through unbroken. Conversely when the lower road is at natural ground level and the higher road is on built up embankment, solid abutments on the edge of the lower road would normally be provided. Between these extreme cases other considerations may prevail, for example where the upper road is so wide that there would be too much of a tunnel if solid abutments were provided; in such cases open spans with columns or pierced walls as piers would be preferable. A similar need arises where visibility is of special importance.

There may be occasions when solid abutments are desirable in order to exploit the heavy character of a bridge.

21. Within the general purpose to be served there will be *engineering requirements* which will largely determine the type of structure. Any detailed account of them would be outside the scope of this booklet.

22. There are also the *physical properties of the site*, the nature and strength of the subsoil, the liability to flooding or subsidence, etc. These again are purely engineering aspects which are not within its scope.

23. It is sometimes thought that when all these factors have been considered, the designer is in a position to decide the general lines of his solution, but full weight should also be given at this stage to what might be called the *aesthetic character of the site*. In a sense this is also a physical aspect, the essential difference being that it does not lend itself to mathematical expression.

24. First there is the general character of the surrounding land, its conformation; flat, undulating, hilly, etc.; and its use: open heath, agriculture, woodland, industrial development, etc. This will be relevant to any distant view of the bridge.

25. Then there is the more specific character of the particular site, its natural features: rocks, trees, etc; and existing man-made additions, especially buildings. Such elements often constitute satisfying features in the setting and regard should be paid to them in the detailed design. (Plates 7 and 8).

26. The following notes indicate some aspects which need to be borne in mind when the appearance of certain types is being considered. As this booklet is concerned only with the appearance of bridges no reference is made to the structural design requirements for any particular type.

Simply supported slabs and beams

27. This type of design, with the usual slight camber, can have a satisfactory appearance as long as the span does not involve too great a depth of beam or slab, which produces a clumsy effect. This may be countered to some extent by reducing the thickness at the edge. Where for practical reasons a solid parapet is unavoidable, it may be preferable for the whole design to express a similar heaviness.

Continuous contruction

28. In a multi-span bridge a lighter appearance can generally be achieved by continuous construction. But where there is a variation in depth care is necessary to ensure a 'clean' line. This matter is more fully discussed in para 83.

Cantilever construction

29. For longer spans cantilever and suspended span construction is often adopted. This form of bridge can have a most attractive appearance, although the end spans in such briges have sometimes been criticized as having too much the appearance of a half-arch. Good proportions and care in the design at the junction with the abutment will help to remove this association with arch construction. In some

9 Bridstow Bridge, Ross on Wye

10 Usk Bridge, Abergavenny

11 Maidenhead Railway Bridge

9

10

11

cases it may be possible to mask the cantilever balancing arm within the abutment, as in Waterloo Bridge. Special care is even more important when the junction is with an approach viaduct, where the effect of a smooth flow is likely to give the best appearance. If this cannot be achieved, it may be better to introduce a variation in size or treatment for the pier to mask or punctuate the transition. (Plate 26).

30. In the case of a cantilever and suspended span bridge it is important to preserve continuity of the soffit curve (Plate 9); otherwise there is danger of an ugly flatness creeping in over the length of the suspended span. This consideration does not, of course, apply to the through-girder type of bridge such as the Forth Railway Bridge. Any noticeable difference in curvature on either side of the pier should generally be avoided, or if unavoidable made obvious.

Portal frames

31. In single spans a shallower depth can be achieved by a portal frame design. Here again, whether the frames are single or multi-bay, great care is required with the profile. An interesting variant of the portal frame design which will give a lighter appearance over long spans is shown in Plate 10.

Arches

32. Spandrel filled arches may be in concrete, with or without masonry or brick facing. Open spandrel arches may be in steel or concrete.

33. The solid arch is sometimes aesthetically desirable in certain settings, for example where a bridge impinges on an old city wall or other solid masonry structure. Some very beautiful arch bridges of the heavier type have been built from the time of the Romans onwards. During the last century Brunel adopted the semi-elliptical arch for many of his structures, with notable success both structurally and aesthetically. (Plate 11).

34. Where a solid arch is constructed in concrete and the vertical faces are finished with stone or brick, these facings should be functional in appearance, particularly those to arch rings. Whereas arch bridges of this character were often enhanced in the past by specially dressed stone courses, quoins and other features, it is seldom desirable or practicable to attempt such elaboration today. Simple string courses and copings are usually sufficient (Plate 12); any further elaboration demands a high standard of design and craftsmanship if it is to achieve a satisfactory appearance.

35. Good design results can now be obtained in reinforced concrete or steel, which enable lighter arches with open spandrels to be constructed over quite long spans (Plate 13); these may also have a low rise-span ratio. There are many attractive varieties of this form of bridge including a heavy arch with light columns. (Plate 14).

36. If a steel arch is used, efforts should be made to preserve the cleanness of the design by the utmost simplification of the spandrel bracing.

37. Other types of steel arch are exemplified in the Volta Bridge in Ghana (Plate 15) and in the Widnes-Runcorn Bridge, where the transverse girders carrying the deck are hung from the arch where it rises above the road. (Plate 16). If a satisfactory appearance is to be achieved the arch should be allowed to dominate the rest of the structure and should not take the form of a bowstring girder, unless exceptional circumstances require this type of design.

Suspension bridges

38. When this form of design is adopted it is generally for very large spans but it can be used with advantage for small ornamental bridges. In all cases, the curved line of the cables is the arresting feature. Particular care is needed to make the most of this beauty of line.

39. If the cables come down so near the deck at the centre as to be below the balustrade level this may spoil the side elevation. The modern practice of exposing the anchor cable below deck level is aesthetically unfortunate if there is a sudden change in direction. (Plate 17).

40. A difficult problem is the design of the towers, Early suspension bridges generally had massive twin towers with an arch between and this was in harmony with the rather heavier design tradition of those days. A lighter appearance is now generally preferred and can be achieved by fairly slender steel or concrete twin towers with a minimum of bracing members between. (Plates 18, 19). The treatment at the tops of the towers also needs careful consideration as these are so conspicuous; non-functional ornamentation here is likely to appear unreal and out of place.

Cable-stayed girder bridges

41. Another form of bridge that makes use of cables is the cable-stayed girder where a continuous steel box section is supported at intermediate points between the piers by inclined cables carried over a tower. While such a design may look satisfactory with single stays this might not be so if there were too many of them. (Plate 20).

Movable bridges

42. The various types of movable bridges—swing, bascule, vertical lift, transporter etc—all present special problems, particularly rolling lift bridges with their unsightly high curved ends. It would be impradticable to generalise about movable bridges other than to stress that their nature and

12 West Woodburn Bridge

13 Taf Fechan Bridge

12

13

14 Rossgraben Bridge, Switzerland

15 Volta Bridge

16 Runcorn-Widnes Bridge

14

15

16

17

18

17 *See para 39*

18 Tamar Bridge

19 Severn Bridge

20 Wye Bridge

19

20

21 Preston Bypass

22 Doncaster Bypass

23 Abergavenny Road Railway Bridge

21

22

23

24 Brodsworth Colliery Railway Bridge

25 Alverley Railway Bridge

the prominent position they often occupy call for special care to achieve a satisfactory solution, appropriate to their setting, in both open and closed positions.

Footbridges

43. The light loading associated with footbridges provides opportunity for slender and elegant design. Where the approaches are from a lower level it is important to keep the steps or ramps as light as possible, particularly when both are required. Effective designs have been achieved when the approach steps and the footbridge itself form a single structure. Full use should be made of camber to achieve headroom and reduce approach steps or ramps. Wherever practicable solid parapets should be avoided. (Plates, 21, 22).

Railway Bridges

44. The need for bridges carrying railways to be erected and maintained with minimum interference with rail traffic limits the number of suitable designs.

45. Steel plate girders are much used for moderate spans and they should be as shallow as engineering requirements permit. (See para 53 below).

46. But they will of necessity be fairly deep, and any impression of heaviness should be reflected also in the design of the abutments to avoid an unbalanced appearance. (Plate 23).

47. Where a parapet is required above the girder, this should be simple and light in character, of small angles or tubes without the introduction of pedestal blocks. Vertical members supporting parapet rails should link up with or be in phase with any stiffeners that may be provided.

48. A difficult part of the design occurs where the top of the wing wall joins the abutment. Protective fencing in the line of the bridge is generally required for a short length at each end of the girders. The provision of masonry or brick parapet walling for this purpose is inclined to look clumsy and a light open parapet is generally preferable. The junction of the abutment, wing wall, girder and fencing needs to be very carefully handled. The detail should be kept as simple as possible, fussiness or over-elaboration being avoided at all costs. (Plate 24).

49. A pier in the central reservation, when it supports a bridge over a dual carriageway should be of simple contruction and should not normally be carried up to break the lines of the superstructure, nor generally need it mask the bearings.. (Plate 25).

Types in combination

50. A word must be added about different types of structure in combination. This may occur, for example, where a long viaduct crosses a moderately wide river. In such a case the physical problem will generally dictate the character of the main bridge and the approaches will differ in scale and form, often contrasting with the main structure. (Plate 26). With narrower crossings the main span may well harmonise with the viaduct approaches preserving a common character throughout. In this matter, as in most others, where there are aesthetic considerations, vagueness, indecision and the weak compromise are the cardinal sins.

51. *Some further examples of interesting modern design are here illustrated, but do not call for particular comment.* (Plates 27-31).

24

25

26 Medway Bridge

27 Bridge near Orly Airport, Paris

28 Footbridge at Southend-on-Sea

29 Bridge at Wuppertal, Germany

26

27

28

29

30 Hammersmith Flyover

31 Beaconhill Bridge

30

31

Alternative materials and their treatment

General

52. Whatever type of bridge it is decided to use, there are certain problems with finished appearance that are more conveniently considered under the headings of the materials used. The choice of material must be determined to some extent by the character of the surroundings. Thus for a masonry faced bridge it is usually better to use natural stone in a stone area in open country or in a completely stone-built town, whereas in a brick area or brick-built town the use of brick facing would be more appropriate. In either case it may be thought preferable on occasion to use exposed concrete or steel and to exploit the contrast. Paras 52-57 should be read in conjunction with Appendix B.

Steel

53. Where plate girders are used, the clean appearance of a welded bridge is generally to be preferred to a riveted structure, and this is sometimes further improved by the omission of stiffeners on the outside face. (Plate 32).

54. It is important that a suitable colour should be chosen for the painting of steel bridges. A uniform colour throughout is usually best. If it is necessary to emphasise one part of the structure, for example to distinguish the main span of a bridge from an approach viaduct where they are not already clearly enough distinguished by their size or form, some contrast in colour may be useful; but this needs very careful handling. The impression of lightness in a parapet can be emphasised if desired by painting it a much paler colour than the rest of the structure.

55. Neutral colours are usually better than bright ones. If there are special considerations which call for inconspicuousness in the landscape, a dark colour will help; if inconspicuousness against the sky is required a lighter colour is better. On the other hand if it is desired to call attention to the bridge, for example to a well designed ornamental footbridge in a public park, a contrast with the background in tone or colour is desirable. White is often very effective, and Chinese red on occasions. Both black and white are best given a very slight bias, black towards blue and white towards cream. More generally, a greyish colour is most useful, since it will go well with both natural surroundings at all seasons and with most ordinary building materials. But here again, an absolutely neutral grey is apt to look dead and dingy. A bias towards blue, green, red or yellow is desirable, the actual choice depending on the surroundings. (Examples from BS 2660: 1955 are 7–076, 7–078, 5–059, 3–035, 3–036 and 2–027). A bluish-grey or reddish-grey is perhaps more appropriate in a sophisticated architectural setting than in open country. The use of a strong green designed to match surrounding foliage is nearly always a mistake.

Other metals

56. Metals other than steel are rarely used, although there may be cases where aluminium is considered for opening bridges or parapets. Wrought iron or bronze are also sometimes used for parapets especially where fine finish or decorative effect is required.

Concrete

57. The heavy appearance of mass concrete, contrasting with lighter members elsewhere in the structure can with advantage be emphasised in the surface treatment. (For illustration see para 59).

58. Reinforced concrete on the other hand with its ability to take tension offers opportunity for exploiting the quality of slenderness. This may be further emphasised in appropriate cases as discussed in para 83 below. Sharpness of definition, conducing to a feeling of crispness is the effect to be aimed at; this will tend to be lost if too much use is made of rounded sections. With the thinner members used in such structures and the added subtlety of line it becomes specially important to ensure that there is no distortion resulting from inadequate design of the bracing or support of the formwork.

59. Smooth concrete surfaces show up any staining and where this cannot be prevented special measures are desirable to render it less conspicuous. Particularly on the broader surfaces, where it is in any case more difficult to produce a perfectly smooth surface, it may be an advantage to provide deliberately a rough texture. Some alternative examples are:

- **(a)** Natural boardmarks from timber shuttering, the degree of roughness depending on circumstances, or the smooth square pattern from steel forms. In all cases good workmanship is essential. (Plate 33).
- **(b)** Shutter Linings, giving a wide range of textures to choose from. (Plate 34).
- **(c)** Exposed aggregates, with the opportunity they afford to introduce pleasing colours, can be achieved by brushing, tooling, grit blasting or other techniques. (Plate 35).

For practical reasons and to preserve sharpness of definition it is generally undesirable to take any rough finish to the extreme edge of a plane surface. (Plate 36).

Masonry: general

60. It is convenient to consider the use of masonry under the two heads of stonework (including artificial stone) and brickwork, since the character of the materials differs so much in appearance and building technique. In either case, they can be used as structural materials or as facing materials.

32 Samlesbury Bridge

33 4 inch × 1 inch sawn timber. Formwork stripped after 2 days (Photograph represents 8 sq ft)

34 Formwork lined with thermoplastic (Photograph represents 8 sq ft)

35 Gap graded concrete $1\frac{1}{2}$ inch to $\frac{3}{4}$ inch beach gravel aggregate (Photograph represents 8 sq ft)

36 Point tooled in-situ Concrete (Photograph represents 8 sq ft)

32

33

34

35

36

Stone

61. For smooth surfaces, particularly where a specially good finish of prime importance, natural stone is usually preferable to artificial stone, which may not weather so satisfactorily. For a rough texture this is less important and good results have been achieved by the use of specially prepared concrete units broken in half and laid so as to expose the broken face. (Plate 37).

62. If natural (or artificial) stone is used merely as a veneer, it is now generally considered better practice not to let it appear to be a structural material, and whatever the backing to form a regular or irregular pattern that cannot be confused with, say, bonded ashlar work. On the other hand a structural stone face with a concrete backing should express true function.

The use of stone is perhaps more suitable in those areas of England and Wales in which it is a traditional building material.

63. A major consideration in the selection of stone is its texture, ranging from smooth polished marble or granite at one extreme to rough granite or limestone, or flint at the other. This is particularly important where the structure is likely to be viewed at close quarters. At a greater distance it is the method of laying which will be more important, ranging from smooth ashlar to random rubble. The pattern of joints in coursed work will also play an effective part. Complicated patterns are usually a mistake in any structure such as a bridge which is not expected to be the object of prolonged contemplation. Smooth surfaces are more appropriate to sophisticated urban surroundings, rougher ones to open country, but this aspect of design is a subtle one and there can be no general rule.

Brick

64. From the point of view of appearance the important variations in the character of brickwork lie in its colour and texture, the size of unit and the thickness of joint. Unless the bridge is being erected in the middle of a red brick area or in association with existing red brick structures, it is usually better to avoid very bright colours. There is now a very wide range of dull yellow, brown, purple or grey bricks which are attractive in themselves and can look particularly well contrasted with some other material or colour in association with them, for example dull yellow with Portland stone, or grey brick with red brick dressings. The use of very rough bricks, particularly those simulating natural roughness, such as rustic Flettons is usually undesirable and unnecessary. Where the scale is suitable the appearance may sometimes be enhanced by the use of bricks below the normal thickness.

Timber

65. In this country there is seldom any call now for permanent timber bridges, except perhaps small footbridges in rural areas or in urban parks. If designed simply and constructed with care they can form attractive features. Softwoods need the protection of paint or varnish, and if designed as ornaments to the scene painted bridges should be white or perhaps some strong primary colour. Hardwood, such as oak or teak is best left unpainted. (Plates 38, 39).

37 Broken pre-cast units

38 Footbridge at Harlow

39 Footbridge at New Sarum

38

39

40 Bridge on M5

41 Winthorpe Bridge

42 Cuerdale Road Bridge, Lancs.

43 Bridge on M5

40

41

42

Detailed design

General

66. If a satisfactory appearance is to be achieved in new bridges, very careful attention must be given to the detailed treatment of the various elements in the structure. Some of those details have already been dealt with, for example surface finishes, which depend directly on the material used for the structure and are therefore more conveniently considered under the general heading of materials. There are others where the same considerations apply whatever the material used and these are discussed in the following paragraphs.

Abutments and wing walls

67. The top of a wing wall should be designed to avoid unnecessary breaks, so as to give a clean and simple effect. If there is a coping its depth should be carefully adjusted to the proportions of the wall and the general character of the bridge. Similar considerations apply to its projection; this is usually best kept substantially less than the depth of the vertical face.

68. The relation between wing walls and abutment is particularly important from the point of view of appearance. In bridges of the lighter type a close harmony in finish and in detailed treatment may be desirable so that both together will form an effective contrast with the light superstructure. In other types of bridge, portal frames for example, the abutment will appear integral with the superstructure, and contrasting wing walls may improve the appearance. (Plate 40).

Piers

69. In many cases, it is an advantage if the form of the piers bears some resemblance to that of the abutments and the finishes are the same. This is obviously essential if the spans take the form of arches, or of beams with appreciably curved soffits. In other cases, for example when a light concrete bridge is supported on heavy abutments but on narrow piers, there is likely to be so great a contrast in any case between the appearance of abutment and pier that it may be better to exploit this contrast in the detailed design and finishes. (Plate 41).

70. Piers may be solid or pierced with openings. Solid piers may be necessary for practical reasons, e.g. in rivers subject to flood and flotsam. Elsewhere, particularly over roads, it is important to reduce the tunnel effect, and, except under relatively narrow bridges, the appearance of the bridge will be improved if the piers are pierced. Openings may take a variety of forms ranging from comparatively narrow apertures kept well up from the ground and well below the soffits to full height openings between narrow pillars. In general it is better to give the pier very clearly one character or the other; either it should be a pierced length of wall or a row of columns. (Plate 42). There is a critical stage in the transition from one to the other which is apt to look ambiguous and for that reason unsatisfactory.

43

71. If the cill-beam is brought above ground level or a continuous plinth is provided, for example to protect the pier from damage, it should appear either as an integral part of a wall pierced with openings or as a base to a row of columns. Again ambiguity should be avoided. The same considerations apply to capping beams, and there is much to be said for similar treatment of cill and capping beam in the same pier.

72. In the pierced wall type of pier a satisfactory appearance will depend mainly on the actual proportions and form of the opening: the ratio of width to height, of width to thickness, of width of opening to width of solid and of height of opening to depth of cill or capping beam. These dimensions are of course determined mainly by structural requirements, but within the limits allowed it is usually possible to arrive at proportions which will appear reasonably satisfactory from varying viewpoints. (Plate 43).

73. It is fairly common practice nowadays for the end faces of such piers to be formed at an angle to the vertical, the pier being longer at its base than at its top or vice versa. Such treatments can often improve the appearance of a bridge provided the slopes are not exaggerated. In general a reduction in the length of the base will give the structure a lighter look, while an increase will make it seem more robust. Devices such as this can add interest to a structure and can give added force to the exploitation of its light or heavy character. Similar variations may also be made in the thickness of the pier.

74. Where the pier consists of a series of columns these are generally best kept vertical, although in some river bridges of

44 Langstone Harbour Bridge

45 Footbridge on M1

46 Wentbridge Bridge

44

45

46

the trestle type the outer piles may be slightly raked for structural reasons and be acceptable in appearance. (Plate 44). Where there are only two supporting columns, these again may sometimes be raked with advantage.

75. The columns can often be improved in appearance by being made polygonal or circular in section. This has obvious advantages with skew bridges, but even in a square bridge it may still serve to lighten its appearance or add interest. Where there is much pedestrian circulation under the bridge, either of these may seem more appropriate than square columns.

76. In some multi-span bridges there may be a change of skew between adjoining spans, involving a pier system wider one end than the other. These cases require special care and skill to ensure a satisfactory appearance.

77. Supports may also be raked longitudinally and many effective structures have been designed in this manner. In such cases the supports may take the form either of solid slabs or of a series of legs. If the legs are also raked transverseley, the amount of rake in this direction should be moderate. (Plates 45, 46).

78. In skew bridges, unless the piers are set well back from the face of the superstructure, the ends should generally be either parallel with the superstructure or polygonal or rounded. (Plate 47).

Capping beams and cills

79. When a capping beam projects beyond the body of the pier or columns it may be desirable to reduce the depth of the end face to lighten its appearance, the underside of the projection being sloped. (Plate 49).

80. With skew bridges, the ends of capping beams usually look better if they are made parallel with the axis of the superstructure, not as in the illustration. (Plate 48). Alternatively, where the pier end is polygonal or rounded the capping beam may sometimes follow the same form. There may be exceptional cases where a heavy capping beam integral with a square ended pier is better square. As far as the cill is concerned, if it is extended with a directional bias, as is sometimes done to give added protection to the pier against damage by traffic, the problem of parallel faces does not arise. Nor of course does it arise in river bridges where the shape of the ends of the cut-waters is determined mainly by hydraulic considerations. Elsewhere it should be related to the pier ends.

Bearings

81. Over piers and abutments a good effect can often be achieved by exposing the bearing, but the treatment should be kept simple. (Plates 50, 51). The emphasis, for example by changes of colour, of individual parts of the bearing or of the base is liable to appear fussy and unsatisfactory. There may be cases, however, where the siting and character of the design or need for uniformity call for masking; this should always be done unobtrusively. In suspended spans the break at the bearing is best left undisguised.

82. There have been cases where the use of steel plates in the bearings of suspended spans has resulted in rust stains on the face of the bridge. This is a serious matter and care should be taken in the design to eliminate this risk. There are of course other features (e.g. metal parapets) which call for special precautions to avoid the risk of unsightly stains.

Face treatment: beams of uniform depth

83. The treatment of face beams has an important effect on the general appearance of a bridge. Where the depth of a beam gives an undesirable appearance of heaviness this may be relieved by avoiding an unbroken vertical face. Changes of plane can usually be combined with variations required for structural reasons, weather protection, services, etc. It is of the utmost importance, however, that the ratios between the depths of the different faces and between depths and projections should be satisfactory to the eye. It is usually best to have a main face to the beam, other vertical faces above or below it being markedly shallower. Projections are usually more effective if the horizontal dimension is kept much smaller (or, more rarely, much greater) than the depth of the vertical faces above or below them.

84. With shallower beams such breaking up of the vertical face should be avoided. Recessed panels with vertical interruptions are seldom satisfactory for any depth of beam. Any form of repeating feature, such as masks to anchorages for transverse post-tensioning, breaks the smooth run of the beam and wherever practicable these should be located behind the face beam. If this cannot be done and a special surface feature is necessary it should be very carefully designed.

85. The use of plastic panels on face beams, contrasting in colour with the parapet, has been successfully adopted on some recent bridges over motorways, but this calls for skilful handling.

86. In continuous construction, where a variation in the depth of the main member is required from span to span or within the span for structural reasons, the appearance can sometimes be improved by chamfering back so as to preserve a uniform depth of vertical face.

Expansion joints

87. Normally a bridge looks better if the expansion joints are undisguised, but care is needed in the detailed treatment.

47

48

47 Chrismill Bridge

48 *See para 80*

49 Bridge on Doncaster Bypass

50 Samlesbury Bridge

51 Bridge on M5

49

50

51

52 Countisbury Bridge

53 Waterloo Bridge

52

53

Faces of bridges with curved soffits

88. Treatment of the faces of true arch bridges raises quite different problems. If a concrete arch ring is faced with stone, the voussoirs should be appropriate in depth to a masonry arch (not as shown on plate 4); many pleasing variations in the detailed treatment have been handed down from classical times. In suitable settings spandrels have been successfully faced in stone with radial joints forming a fan-shaped pattern above the ring. (Plate 52). With brickwork, however, the smaller scale and regularity makes such a treatment unsatisfactory (see also para 34).

89. With other forms of construction involving curvature in the soffit, should be decided in the exceptional case to clad the vertical face, it is probably better to emphasise that this is a veneer by introducing vertical jointing or in some other way that avoids the suggestion of voussoirs, the bottom margin being finished as a continuous face of narrow and even depth, possibly of contrasting colour, slightly projecting, recessed of chamfered. (Plate 53). (See also para 62).

Soffits

90. Although the undersides of most modern bridges get only a fleeting glance from the passing motorist, there are some, for example over rivers or in built-up areas, where there is opportunity for closer observation. In such cases particular attention should be paid to the finished appearance. In wide bridges for example, unless they are masonry arches in which voussoirs form their own pattern, it is usually desirable to provide some relief, if only by a special texture or by small chamfers or V-joints between beams laid side by side. (Plate 54). Relief is provided naturally where a road is carried on twin bridges or where there is a light well.

Subway linings

91. In pedestrian subways the first consideration must be light reflection. Walls should be finished so that they cannot be written on, and they should not be used for advertisements. Fairly neutral colours are better than very bright ones. Fussiness in detail should be avoided.

Parapets

92. The design of the parapet is among the most important factors determining the appearance of a bridge. Although it may sometimes be necessary to have a solid parapet for practical reasons, or occasionally even for aesthetic reasons, for example to give added depth and weight above an arch, it is usually best to keep the parapets light and open. Some simple combination of posts and rails looks well on most modern bridges. (See para 82).

93. There are two common forms depending on the relative emphasis of vertical or horizontal treatment. The horizontal treatment in which three or four rails are carried on widely spaced standards is appropriate for a bridge carrying a fast uninterrupted flow of traffic, for example a motorway. (Plate 55).

94. On other bridges, particularly those providing for pedestrians, where the need for expressing uninterrupted flow is not a factor, the vertical treatment may be appropriate, with bars at not more than 5″ centres. (Wider spacing involves the risk of childrens head's being trapped). This is also a safer type where children are concerned. (Plate 56).

95. There is also the case of the protective fence separating footpaths and cycle tracks from a motorway when both are carried on the same bridge structure (as at Maidenhead and Medway), or separating dual carriageways. This type of fence has particular safety functions; the aim should be to keep the design as simple and unobtrusive as possible and plain vertical bars with a single toprail generally serve best. (Plate 57).

96. If for some special reason, for example in a city centre, the design of a bridge is required to be more monumental, the treatment of the open parapet may be elaborated by emphasising the main verticals, by the introduction of embrasures over projecting piers or by some other bold relief. The introduction of ornamental panels or solid dies or blocks is apt to appear fussy and unsatisfactory. If commemorative plaques are to be added provision for them should be made in the original design.

97. Bridges over railways present special problems arising out of the need for construction, maintenance and renewal with the minimum interference with rail traffic, and also on account of safety precautions, including Ministry of Transport Regulations. Parapets in such cases are designed to protect the railways as well as the pedestrian, and requirements are more exacting where there is overhead electrification. The parapet, or at least the lower part of it, must be solid, and the inner face must not provide a foothold for climbing. The normal height is 5′ (4′ if the line is not electrified). There are special provisions for motorways. In exceptional circumstances a concrete canopy has been agreed as an alternative to a high solid parapet.

98. The difficulty of making a railway bridge parapet both safe and acceptable in appearance is even greater when the bridge is part of a lengthy viaduct. To make the whole length of parapet correspond to the portion over the railway is generally objectionable and the alternative of making a clean break, for instance from a 5′ solid parapet to a light open one 3′ 6″ high, may, be preferable. When the whole length is built in brick it may sometimes be acceptable to step down the height in one or two stages on either end of the portion spanning the railway.

54

55

54 Chiswick Flyover

55 Chiswick Flyover

56 Sittingbourne Road Bridge

57 Thames Bridge M4

56

57

58 Lanercost Bridge

59 Berwick Bridge

58

59

Problems of historical bridges

99. With the increasing volume and weight of traffic more and more of the country's older bridges are proving inadequate, either on account of structural weakness or because they are too narrow. In many cases they can be strengthened or widened or completely rebuilt without raising other than structural or economic questions, and of course the normal aesthetic problems of finished appearance already discussed in this booklet. There are other cases, however, where the existing structure is of such outstanding architectural or historical importance that any proposal to demolish or alter it raises questions of a different order. Sometimes, indeed, the structure may have been scheduled as an Ancient Monument by the Ministry of Public Building and Works. More often it may have been listed by the Ministry of Housing and Local Government under the Town and Country Planning Act of 1947 as of special architectural or historical interest. Many bridges which have not been so listed may also be of such intrinsic interest from this point of view that a strong case can be established for their preservation or, if alterations or additions have to be accepted, for these being so designed as to interfere as little as possible with their existing character.

100. An old bridge may be of such outstanding importance as an historical monument that neither its demolition nor any material alteration to its appearance should be accepted. The difficulties may be overcome by the construction of a by-pass, and it should be emphasised that where this is contemplated within a reasonable time, every effort should be made to avoid disturbing the old bridge. In other cases it may be necessary to build a new bridge alongside. (Plates 58, 59). The existing bridge may still have to be repaired if it is not to be allowed to collapse, and it may continue in use as a footbridge.

101. If the disadvantage of the existing bridge is its narrowness rather than its structural weakness, it may sometimes be possible to use it with or without strengthening as a one-way bridge. In such a case also it will be necessary to build a second bridge alongside, unless some radical change in traffic circulation provides an alternative solution. Again the width of the bridge may be adequate for two-way traffic if existing pedestrian traffic is eliminated, and for this purpose a pedestrian footbridge alongside may be all that is required.

102. Where a new bridge has to be built close to an existing one it is important that it should not be too near. The main reason for preserving an existing bridge without alteration in spite of its inadequacy is to enable people to enjoy the view of it in its setting: if a new bridge is built immediately alongside this enjoyment is lost, certainly as seen from one side and quite possibly from both. Even at some extra cost it is usually well worth while keeping a reasonable distance away and diverting the approaches as may be necessary. (Plate 58). Beyond this it is clearly impossible to generalise. Each case will require very careful and detailed examination. If the two bridges must be close to one another there is generally an aesthetic advantage is making them parallel. (Plates 59,60).

103. In the design of a new bridge to be erected near an old one it is usually a mistake to copy the old work. An exception might be the case in which an existing bridge of fairly conventional design, say a small 18th century bridge in an urban setting, is to be duplicated for one-way traffic.

104. Normally, the new bridge should be a good example of its own period, acting perhaps as a foil or contrast to the old bridge, whether or not similar materials are used. The contrast between old work of massive construction and new work of concrete or steel can often be effectively exploited. (Plate 61).

105. The problem is different in the case of a new bridge dominating its older neighbour by sheer size, and that of a light footbridge forming little more than an appendage to the existing bridge. In the former case keeping a reasonable distance becomes the more important factor although relative levels may sometimes require equal consideration; in the lattercase while distance is still important the detailed design and the element of contrast are what matter most. (Plate 62).

106. In this connection the reader is referred to what has already been said in this booklet about footbridges; in addition there is the possible use of timber bridges as suitable accompaniments to ancient stone bridges, to which timber seems a natural foil.

107. In those cases where it is accepted that an old bridge may be altered it may still be important to ensure that such alterations are carred out with minimum change in outward appearance. Such changes as may have to be made should generally incline towards simplicity rather than elaboration.

108. Where it is necessary to widen an old bridge, it is sometimes assumed that provided the elevational treatment is kept the same nothing else matters. This is not necessarily so. The pictorial effect of an old bridge may well depend upon its width seen through an arch, the amount of sky visible obliquely and the amount of scenery visible below the sky. For this reason it may often be better aesthetically to keep the existing arch or arches as they are and cantilever out for additional width.

109. If cantilevering is accepted it is more likely to succeed with the lighter classical type of masonry bridge than with the heavier medieval type in which the cutwaters are

60 Tamar Bridge

61 Lanercost Bridge

62 Giggleswick Bridge

63 Brecon Bridge

64 Bridgnorth Bridge

60

61

62

63

64

carried up between the arches to form embrasures in a continuous solid parapet. Here widening the whole structure and refacing may be the better answer, but it is sometimes possible to sacrifice the embrasures in order to run a light footpath across the top of the cutwaters, keeping a solid parapet on the inside. (See also para 112 below.) (Plate 63).

110. If the cutwaters are not carried up, and in the classical type of bridge they are usually stopped at or near the springing of the arch, it may sometimes be possible and aesthetically acceptable to carry the footpaths on longitudinal beams with supports based on the cutwaters (Plate 64). More usually, however, it will be better to provide the extra width by a cantilevered slab or a series of brackets.

111. The satisfactory design and the placing of these brackets raise problems of exceptional difficulty from the point of view of finished appearance. The situation varies so much with different bridges that no general principle can be laid down, but the aim should be to continue as far as possible the logical development that lies at the root of all classical design. If, for example, there is a full entablature above the arch the transverse brackets should be inserted in the frieze if possible, the architrave being left intact and the cornice reproduced on the new outer edge. (Plates 65, 66).

112. The treatment of the parapet over the cantilevered portion is another difficulty. A classical stone balustrade, for example, that looks satisfactory in the plane of the spandrels may well look too heavy if brought forward and carried on brackets, as well as being structurally undesirable. It may be better to substitute a lighter parapet altogether. For example, in a masonry bridge of great apparent weight and bold profiles, the old parapets can sometimes be left in situ dividing the carriageway from the footways, the lightest type of metal railing being erected on the new outer edges.

113. Special consideration must be given to the ends of the parapet in this sort of cantilevered widening. Very often the ends of the original bridge will have been curved round to form a smooth junction with the embankment on either side or possibly a reverse quadrant will have been inserted between the two to punctuate the transition, in which case the old parapet will follow the line of the structure below, while the new one will either cut across this concave or convex curve or occupy its full width or even extend beyond it. In all such cases the detailed design of the widening, especially as seen from below, calls for very careful study if the effect is not to appear makeshift. (Plate 67).

114. Mention may here be made of alterations to bridge approaches. Whether an existing approach is to be widened on one or both sides, or whether it is to be supplemented by separate approach to a new bridge alongside, special attention should be paid to the treatment of any new embankments and retaining walls, to the layout and treatment of any small open space between the old and new work. whether paved or planted, and to the connection between the parapet of the bridge and that of the embankment. Even if the new work is designed to contrast to some extent with the old, a proper junction between the two is essential. Confusion such as that illustrated should be avoided. (Plate 68).

65

65 London Bridge

66 Newton Cap Bridge

67 Coldstream Bridge

68 *See para 114*

66

67

68

69 Perth Bridge

70 Waterloo Bridge

71 Hanger Lane Underpass

69

70

Lighting of bridges

115. Except on bridges of very small span any street lighting equipment to be provided should be considered aesthetically as an integral part of the design. Any bridge carrying a major traffic route in an area to be lit will probably need Group A lighting with lamps mounted at 25′ according to the current Code of Practice (1962). If the standards are placed at a short distance behind the kerb on the approaches, this may well be accepted over new bridges, especially where there is no major break in what the road user sees. If there is such a break or if the lamp standards are going to be a conspicuous feature in side views of the bridge, it may be better to site them on a central reservation, or to incorporate them as special features in the parapet. With this latter treatment, however, the longer brackets which may be required to achieve the normal standard of illumination may raise difficult design problems; alternatively the reduction in lighting efficiency may be offset by increasing the number of lamps, for example by arranging them in opposite pairs instead of staggering them. (Plate 69).
116. The lighting of old bridges raises different problems. There are cases where the widening of an old bridge is required as part of a traffic improvement and a new lighting installation is needed at the same time. Most old bridges even when widened are spoiled if lamp standards of 25′ or more are erected on them, except perhaps when they are in a central reservation. The possibility of shorter columns at closer intervals may well be considered in such cases, but the placing of any columns on the outer edge of the bridge should obviously be carefully related to the spacing of piers and abutments. Occasionally, if the overall length of the bridge is not too great it may be desirable to light from each end only, with an increased mounting height and stronger lamps.
117. A word should be added on lighting below bridges. The need for such lighting has in the past been too often ignored in the design of the bridge, but such provision may be necessary—it will be essential in the case of lengthy underpasses—and it should be done in such a way as to produce a good appearance both by day and by night. If the surface of the structure is to be used for reflecting light, this should be borne in mind in considering its finish. (Plate 71).
118. Special lighting problems arise in connection with lights on bridges over railways and navigational lights on rivers. Lamps on bridges over railways often have to be masked in order not to confuse train drivers. This may make the lantern look peculiarly clumsy in the daytime. Any measures that can be taken therefore through careful planning of the installation to reduce such masking to a minimum will be well worth while.

71

119. Navigational lights are often completely ignored in the design of a new bridge and appear as an ill-considered afterthought; most of the Thames bridges in London suffer in this way. Such eyesores should be avoided by collaboration at the outset between the authorities concerned. (Plate 70).

Collaboration between engineers and architects

120. In former times there was no rigid distinction between engineers and architects. The sustained development however of applied science, and particularly the introduction first of iron as a structural material and then of steel and reinforced concrete, has today made it difficult for any individual to excel in every aspect of bridge design.
121. There will be many cases where there is clear advantage to be gained by securing experienced architectural collaboration in the design of new bridges and major alterations to old ones.
122. There may still be a lingering tendency to think of the architectural contribution to bridge design as being something added after the structural design has been completed. This approach to the problem has produced some unfortunate results. If an architect is brought in, full advantage should be taken of his special training and experience from the outset, i.e. when considering any practicable variations in siting, the type of bridge and the materials to be used, as well as the final form and appearance of the structure.

Appendix A

Role of the Royal Fine Art Commission

1. The Royal Fine Art Commission was appointed in 1924 'to inquire into such questions of public amenity or of artistic importance as may be referred to them from time to time by any of our Departments of State, and to report thereon to such Department; and furthermore to give advice on similar questions when so requested by public or quasi-public bodies, where it appears to the said Commission that their assistance would be advantageous'. These powers have been extended to allow the Commission to take the initiative, but it remains essentially an advisory body, with members, including an engineer as well as architects, chosen for their individual abilities and appointed by the Sovereign on the advice of the Prime Minister. The Commissioners give their time voluntarily and are unpaid.

2. From its inception the Commission has given the Ministry of Transport and local Highway Authorities its views on designs for many new bridges or major alterations to old ones and it is generally felt that this collaboration has helped to improve or maintain the standard of appearance in such structures.

3. Both the Ministry and the Commission wish this work to continue; at the same time by its nature the Commission cannot undertake an unlimited amount of advisory work. Bridge design is only one of many subjects with which it is concerned. It is hoped that the publication of this booklet, based as it is on nearly 40 years experience of collaboration between Ministry and Commission, may assist in the preparation of a greater proportion of satisfactory designs and in time appreciably reduce the need for consultation. The Commission's power remains unaffected and it will still expect to be consulted where large schemes or structures are involved, or where there are special historical, architectural or scenic aspects, or peculiar difficulties and strong differences of opinion. The position will be reviewed from time to time with the Ministry.

Submissions to the Commission

4. It has been found more satisfactory for submissions to the Commission to be made through the Ministry of Transport rather than direct by Local Authorities. The Ministry's approval is generally required in any event and this procedure avoids the possibility of conflicting criticism, since a representative of the Ministry is present when schemes are considered by the Commission's Examination Committee. If an Ancient Monument is involved, the Ministry of Public Building and Works should be notified at the outset and they will be brought into the discussion with the Commission before any final decision is reached. The Examination Committee reports to the full Commission at its monthly meeting and the Commission's views on submissions are given to the Ministry.

5. There is nothing, however, to prevent schemes which do not require the approval of the Ministry of Transport being submitted direct, and they are dealt with in the same way by the Examination Committee.

6. The engineer responsible or his representative, and the architect if there is one or his representative, are normally invited to meet the Committee for an informal discussion of the scheme. Such discussion proves most valuable in making clear any criticism or comments the Committee may wish to make which are not always easy to explain in correspondence. At the same time, it is not the intention of the Commission to assume the role of a consultant by putting forward its own designs or details.

7. It may be useful to mention here that submissions made to the Commission either through the Ministry or direct should include site plans showing clearly the relation of the scheme to the surrounding district, photographs when relevant, plans, sections and elevations. Accurate perspective sketches are also very useful but over-elaborate pictorial effects are unnecessary. A good standard of draughtsmanship is expected. In the more important or difficult cases models may be desirable. The Commission is always interested in the finishes proposed and on occasion sample materials may be necessary to explain the intention.

8. In some cases submissions are made when the contract is about to be let or after a Statutory Order has been made. This gives rise to unnecessary difficulties if any change is recommended by the Commission. Submissions must be made in ample time if the Commission is to fulfil its functions effectively.

The position in Scotland and Northern Ireland

9. The Royal Fine Art Commission is concerned with work in England and Wales. There is a separate Commission for Scotland with offices in Edinburgh. Although the two Commissions work on very similar lines, they are completely independent of one another, and what has been said in this booklet should not be taken as necessarily expressing the views of the Scottish Commission.

10. There is no parallel organisation in Northern Ireland and it has been agreed that the Royal Fine Art Commission should advise the Northern Ireland authorities on bridge design in important cases.

Appendix B

Colour and texture in bridge design

General

1. Pattern and colour should be used on bridges with caution, especially if they take the form of applied decoration.
2. Concrete bridges and steel bridges call for different approaches.

Colour of paintwork on steel

3. The prime function of paint on steel is to afford protection to the metal. There are few places where a contrast of colours is desirable, a single colour or a group of related tones of the same colour being generally preferable. Very occasionally, where the design of the bridge is good enough and the landscape can take it, a single bold colour is successful. But more often it will be wise to relate rather than contrast bridge and background. Strong shades of green clash with vegetation and should be avoided. 'Metallic' colours from bronze-green to copper and silver grey can look well on steel structures.
4. On concrete bridges metal balustrades, even though they may sometimes introduce a note of gaiety, should be painted in colours that are related to the neutral and warm greys of the concrete surfaces. If in doubt a mock-up length of handrailing in wood can be set up on the site and painted with alternative colour.
5. On steel bridges the colour of the balustrades should similarly be related to the 'metallic' colour selected for the outside faces of the main steelwork.

Colour and texture of concrete

6. Concrete being a plastic material lends itself to shapes which, if well executed, do not need embelishment. Where however a single design is to be repeated many times as on a motorway, occasional variations in the treatment of concrete pattern and in parapet colour can help to obviate monotony. But the use of fancy patterns or bold colours is seldom successful.
7. Variation and contrast of tone can often be of more lasting interest than a splash of colour. Monochrome schemes making use of warm and cool neutral shade in tones from black or charcoal to white or off-white look very well provided the contrasts are sharp and definite. The concrete colour itself can be made part of such a scheme by the use of dark and light aggregate, black, normal and white cement and various combinations of these. Normally the light tones being more prominent are used for edge beams and fascias, the recessed members, like the columns supporting the deck, being kept dark in contrast. The use of plastic fascias has no aesthetic advantage and is best avoided. Where there are strong technical reasons for their use the colour should be carefully related to that of the rest of the structure.
8. Concrete surfaces usually require vertical joints between one pour and the next and often horizontal joints as well between one lift and the next. These joints give a basic grid. The practicable intervals between them produce a convincing and appropriate scale. The position of both vertical and horizontal lines can usually be adjusted to make a balanced system. The joints may be made almost invisible by taping between the shutters, or they may be emphasized by being given a deep 'v'. The vertical may be suppressed and the horizontal exaggerated or vice versa as appropriate to the particular circumstances. If precast facing panels are to be used the jointing of the panels should bear some relation in lay-out and size to the natural basic grid referred to earlier. Recessed joints normally give an impression of greater strength than raised joints.
9. The most satisfactory textures in concrete are those which in common with natural materials have a small-scale and irregular pattern raised on the surface, like split, sawn or tooled stone or even timber. The least satisfactory are patterns that appear to have been stamped into the surface by a machine in regular geometrical shapes.
10. On the fascias and other vertical surfaces exposed to the weather, fluting or vertical tooling will help to canalise the flow of rain water and thus provide a pattern which will be emphasized rather than spoiled by weathering.

Acknowledgments

2 *Photograph by courtesy of the County Surveyor of Denbighshire*

3 *Photograph by courtesy of Valentine & Sons Ltd*

6 *Photograph by courtesy of R. Travers Morgan and Partners*

11 *Photograph by courtesy of British Railways Board*

12 *Photograph by courtesy of the County Surveyor of Northumberland*

13 *Photograph by courtesy of Stanley Travers,* AIBP, ARPS

14 *Photograph by courtesy of the Cement and Concrete Association*

15 *Photograph by courtesy of Freeman Fox and Partners*

19 & 20 *Photographs by courtesy of Hylton Warner & Co Ltd*

23 *Photograph by courtesy of British Railways Board*

26 *Photograph by courtesy of Freeman Fox and Partners*

27 *Photograph by courtesy of the Cement and Concrete Association*

28 *Photograph by courtesy of the Borough Engineer of Southend-on-Sea*

33 *Photograph by courtesy of the Cement and Concrete Association*

34 *Photograph by courtesy of the Cement and Concrete Association*

35 *Photograph by courtesy of the Cement and Concrete Association*

36 *Photograph by courtesy of the Cement and Concrete Association*

39 *Photograph by courtesy of F. Futcher & Son*

41 *Photograph by courtesy of W. E. Middleton & Son, Ltd*

61 *Photograph by courtesy of the County Surveyor of Cumberland*

69 *Photograph by courtesy of the Perthshire Advertiser*

Dd.152420 K 40 8/69

Printed in England for Her Majesty's Stationery Office by F. Mildner & Sons, London, E.C.2